BASICS OF PASTURE MANAGEMENT

Maximizing Yields, Sustainability, And Livestock Health

Jasper Mark S.I.

Table of Contents

CHAPTER ONE 3
   INTRODUCTION TO PASTURE MANAGEMENT 3
CHAPTER TWO 9
   IMPORTANCE OF PASTURE HEALTH 9
   KINDS OF PASTURES AND FORAGES 14
CHAPTER THREE 20
   SOIL HEALTH AND FERTILITY 20
CHAPTER FOUR 26
   GRAZING SYSTEMS AND PROCEDURES................................26
CHAPTER FIVE...33
   PASTURE ESTABLISHMENT AND RENOVATION 33
   WEED AND PEST CONTROL 39
CHAPTER SIX..45
   WATERING SYSTEMS 45
CHAPTER SEVEN 52
   PASTURE MONITORING AND ASSESSMENT 52
THE END 58

CHAPTER ONE

INTRODUCTION TO PASTURE MANAGEMENT

Pasture management is the art and science of supervising and working on the health, efficiency, and sustainability of meadows or pastures. It plays an urgent role in the livestock business as well as in environmental protection. Compelling pasture management guarantees that pastures are used proficiently, promoting the development of nutritious forage for animals while minimizing soil erosion and corruption.

At its core, pasture management includes a progression of practices and choices aimed at upgrading the utilization of accessible land assets. This incorporates rotational grazing, soil fertility management, weed control, and water management. Every one of these parts contributes to the general health and efficiency of the pasture ecosystem.

Rotational Grazing

Rotational grazing is a vital part of pasture management. It includes splitting the pasture into smaller paddocks and rotating livestock between them at customary intervals. This practice permits forage to recuperate and regrow in grazed regions while guaranteeing that animals approach new, top-notch grass. Rotational grazing augments forage use as well as assists in decreasing dirtying compaction and promoting plant variety.

Soil Fertility Management

Keeping up with soil fertility is significant for the drawn-out efficiency of pastures. Soil testing and analysis are fundamental devices in determining the nutrient content of the dirt and distinguishing any deficiencies. In view of these discoveries, suitable preparation systems can be carried out to renew nutrients and promote healthy plant development. Organic alterations, like compost and manure, can likewise be utilized to further develop soil construction and fertility.

Weed Control

Weeds can contend with positive forage species for daylight, nutrients, and water, in this manner decreasing pasture efficiency. Successful weed control systems, including cutting, herbicide application, and manual expulsion, are important to hold weed populations under tight restraints. Incorporated weed management draws near, and joining cultural, mechanical, and chemical control techniques can assist with maintaining a balanced and different pasture ecosystem.

Water Management

Appropriate water management is fundamental for both plant development and animal health. Pastures ought to be intended to catch and disseminate water productively, minimizing overflow and erosion. This might include introducing watering tanks, developing lakes or dams, and carrying out seepage systems to upgrade water use and accessibility.

Note that pasture management is a multi-layered discipline that means improving the efficiency and sustainability of prairies for livestock creation. By utilizing practices like rotational grazing, soil fertility management, weed control, and water management, farmers and land managers can guarantee that pastures stay healthy, useful, and resistant to environmental difficulties. Taking on sound pasture management practices not only advantages livestock producers by further developing forage quality and quantity, but additionally adds to environmental preservation by protecting soil health and biodiversity.

CHAPTER TWO

IMPORTANCE OF PASTURE HEALTH

Pasture health is significant for both the environment and the livestock that rely on it. Basically, a healthy pasture is an establishment for practical farming and a flourishing ecosystem. How about we figure out why pasture health is so significant?

A healthy pasture, most importantly, gives nutritious forage to livestock. Grazing animals, like cattle, sheep, and goats, depend on pastures for their food. At the point when pastures are all around managed and healthy, they produce top-notch grasses and legumes that are rich in fundamental nutrients.

This guarantees that livestock get a balanced diet, which is fundamental for their development, proliferation, and, generally speaking, prosperity. Healthy pastures likewise contribute to better animal health, lessening the requirement for supplemental feeds and veterinary interventions.

Besides, pasture health plays a significant role in soil protection and erosion control. Healthy pastures have a thick front of grasses and plants that shield the dirt from erosion brought about by wind and water. The root systems of these plants assist with restricting the dirt together, making it more impervious to erosion. In addition, the organic matter created by healthy pastures further develops soil structure, dampness maintenance, and fertility. This benefits the actual pasture as well as improving the efficiency and resilience of the whole farming system.

Besides, healthy pastures contribute to biodiversity and habitat preservation. A different range of plants in a pasture gives food and shelter to different wildlife species, including bugs, birds, and small mammals. This promotes natural equilibrium and supports a dynamic ecosystem. Pasture management practices that focus on biodiversity and habitat protection assist with saving local plant species and wildlife habitats, adding to the preservation of neighborhood ecosystems and biodiversity.

Additionally, healthy pastures play a role in carbon sequestration and climate change moderation. Prairies and pastures can catch and store carbon dioxide from the climate through photosynthesis. Healthy pastures with vigorous plant development and profound root systems sequester more carbon, assisting with balancing greenhouse gas emanations and alleviating the impacts of climate change. Reasonable pasture management practices, for example, rotational grazing and the utilization of cover crops, can upgrade carbon capacity in soils and add to climate resilience.

Notwithstanding, pasture health is of vital significance for manageable agriculture, environmental preservation, and animal government assistance. A healthy pasture gives nutritious forage to livestock, upholds soil protection and erosion control, upgrades biodiversity and habitat preservation, and adds to carbon sequestration and climate change moderation. Thusly, farmers and land managers ought to focus on pasture health in their management practices to guarantee the efficiency, resilience, and sustainability of their farming systems. By putting resources into pasture health, we can establish a healthier environment, support flourishing ecosystems, and produce top-notch food for people and animals.

KINDS OF PASTURES AND FORAGES

Pastures and forages are imperative parts of livestock farming and play a significant role in giving nutrition to animals. While the two terms are frequently utilized reciprocally, they have particular implications and fill various needs in livestock creation.

Pastures

Pastures allude to areas of land covered with grasses, legumes, or different plants that are reasonable for grazing livestock. These are typically managed to provide a nonstop wellspring of forage for animals. Pastures can be ordered into various sorts in light of the kinds of plants they contain and their management practices:

1. Natural Pastures: Local grasses and plants develop without human interference. They are often found in districts with appropriate climatic circumstances and soil types for grass development.

2. Improved Pastures: These are pastures that have been planted with chosen grass and vegetable species to work on their nutritive worth and efficiency. Further developed pastures are typically managed all the more seriously and can give better returns on quality forage.

3. Rotational Pastures: In this framework, animals are moved between various paddocks or segments of pasture to consider the regrowth of the grass and to forestall overgrazing. This aide in keeping up with the efficiency and health of the pasture after some time.

Forages

Forages, then again, are plants or plant parts (like leaves and stems) that are utilized as feed for livestock. Forages can be developed explicitly for feeding animals or can be collected from pastures, glades, or developed fields. They can be characterized into various kinds in light of their developmental propensities and nutritional content:

1. Grasses: Grasses like ryegrass, fescue, and bermudagrass are ordinarily utilized as forages. They are rich in carbs and fiber, making them appropriate for ruminant animals like cattle and sheep.

2. Legumes: Legumes like clover, horse feed, and trefoil are high in protein and minerals, making them important forage crops. They can additionally fix nitrogen from the environment, which can further develop soil fertility.

3. Silages: Silage is a kind of forage that is matured and put away under anaerobic circumstances. It is frequently produced using grasses or legumes and is utilized as a feed for livestock during periods when new forage isn't free.

Differences Between Pastures And Forages

While the two pastures and forages are significant for feeding livestock, they contrast in more ways than one:

• Origin: Pastures are areas of land covered with reasonable plants for grazing, while forages are the plants or plant parts utilized as feed for animals.

• Reason: Pastures act as a nonstop wellspring of forage for grazing animals, though forages can be developed explicitly for feeding or reaped from different sources.

• Management: Pastures require continuous management to keep up with their efficiency and health, while forages might require development, collecting, and capacity practices to guarantee quality feed for animals.

Notwithstanding, while pastures and forages are firmly related and frequently utilized together in livestock farming, they share unmistakable parts and characteristics. Pastures provide the actual space and environment for animals to brush, while forages supply the fundamental nutrients and energy required for their development and creation. Legitimate management and determination of the two pastures and forages are fundamental for improving livestock nutrition and, in general, farm efficiency.

CHAPTER THREE
SOIL HEALTH AND FERTILITY

Soil health and fertility are urgent parts of agriculture and play a huge role in deciding crop yield, plant development, and overall ecosystem balance. Basically, healthy soil implies healthy plants, which thusly implies healthy food and a healthy environment for us all. However, what precisely do we mean by soil health and fertility?

Soil health alludes to the general well-being and quality of soil, including its physical, chemical, and biological properties. A healthy soil is one that is very organized, very depleted, rich in organic matter, and overflowing with useful microorganisms. These microorganisms play a fundamental role in nutrient cycling, sickness concealment, and soil structure improvement. At the point when soil is healthy, it gives a steady environment to establish roots, permitting them to successfully get to fundamental nutrients and water more.

Then again, soil fertility alludes explicitly to the dirt's capacity to furnish plants with the nutrients they need to develop and flourish. Ripe soil contains sufficient degrees of fundamental nutrients like nitrogen (N), phosphorus (P), and potassium (K), as well as micronutrients like calcium, magnesium, and sulfur. These nutrients are urgent for plant development, blooming, fruiting, and general plant health.

Keeping up with soil health and fertility requires a proactive approach to soil management. Here are a few key practices that can help improve and keep up with soil health and fertility:

1. Crop Rotation: Rotating crops helps break bug and sickness cycles, further develops soil construction, and improves nutrient accessibility. Various crops have different nutrient necessities and contribute various buildups to the dirt, which keeps a balanced nutrient profile.

2. Cover Cropping: Establishing cover crops like clover, rye, or vetch during

decrepit periods or between cash crops can assist with forestalling soil erosion, further develop soil design, and add organic matter to the dirt when they are integrated or mulched.

3. Organic Matter Addition: Integrating organic matter into the dirt using compost, manure, or green manures can increment soil fertility, further develop water maintenance, and animate gainful microbial activity.

4. Reduced Culturing: Minimizing soil unsettling influence through diminished culturing or no-till farming practices can assist with safeguarding soil structure, decreasing erosion, and rationing soil dampness and organic matter.

5. Soil Testing: Standard soil testing is fundamental for assessing nutrient levels and pH, distinguishing nutrient deficiencies or awkward natures, and directing manure and lime application rates to address crop issues without over-application.

6. Balanced Treatment: Applying composts in a balanced and designated way based on soil test results and crop nutrient necessities keeps up with soil fertility without causing nutrient lopsided characteristics or environmental contamination.

7. Integrated Irritation Management (IPM): Carrying out IPM methodologies that focus on forestalling bug issues through cultural practices, biological

control, and designated pesticide use can assist with diminishing the requirement for chemical information sources and safeguarding advantageous soil organic entities.

In any case, soil health and fertility are central to feasible agriculture and food creation. By embracing practices that promote soil health and fertility, farmers can further develop crop yields, lessen input costs, safeguard the environment, and contribute to long-haul agricultural sustainability. Putting resources into soil health and fertility isn't just really great for the dirt and plants; it is also fundamental for guaranteeing food security, environmental health, and the well-being of people in the future.

CHAPTER FOUR

GRAZING SYSTEMS AND PROCEDURES

Grazing systems and procedures play a significant role in pasture management to guarantee economical forage creation, keep up with soil health, and improve livestock execution. Legitimate grazing management can prompt better pasture quality, an expanded conveying limit, and upgraded environmental protection. Here, we'll examine some normal grazing systems and procedures utilized in pasture management.

Rotational Grazing

Rotational grazing includes separating a pasture into smaller paddocks and rotating livestock through these paddocks at customary intervals. This framework permits forage plants to recuperate between grazing periods, prompting healthier and more useful pastures. Rotational grazing can be managed in different ways, including:

• Time-Controlled Grazing: Livestock are moved to another enclosure after a set, not entirely set in stone by forage development and livestock interest.

• Stocking Density: The quantity of animals per unit region is acclimated to match forage accessibility, guaranteeing neither undergrazing nor overgrazing happens.

Continuous Grazing

In continuous grazing, livestock have unlimited access to a pasture for a drawn-out period. While this system requires less concentrated management than rotational grazing, it can prompt lopsided forage usage, diminished pasture efficiency, and soil corruption while possibly not being carefully managed.

Adaptive Multi-Paddock (AMP) Grazing

AMP grazing consolidates components of rotational and consistent grazing. Livestock are pivoted through a progression of smaller paddocks, yet the span of stay and grazing force can be changed in view of forage development and pasture conditions. This adaptable methodology takes into account ideal forage usage and further develops soil health.

Strip Grazing

Strip grazing includes dispensing a limited piece of pasture to livestock, which is continuously moved or moved to give new forage. This strategy is particularly valuable for focused energy and short-span grazing and can assist with boosting forage admission and limiting squander.

Key Grazing Strategies

1. Rest and Recuperation: Permitting forage plants a sufficient chance to

recuperate between grazing occasions is fundamental for keeping up with pasture health and efficiency. Rest periods ought to be customized to forage development rates and environmental circumstances.

2. Graze Level Management: Keeping an ideal grazing level can assist with promoting incredible forage regrowth and preventing harm to established crowns. Grazing excessively near the ground can debilitate forage plants and diminish long-haul efficiency.

3. Water and Shade Arrangement: Guaranteeing simple access to clean water and shade can assist with minimizing pressure and further

developing livestock execution, particularly during blistering climates.

4. Fencing and Infrastructure: Very well planned fencing and infrastructure are fundamental for compelling pasture management. They assist with controlling livestock development, work with grazing rotations, and shield touchy regions from overgrazing.

Note

Powerful grazing systems and procedures are crucial for reasonable pasture management. Rotational grazing, nonstop grazing, versatile multi-enclosure grazing, and strip grazing are some normal grazing systems that can be adjusted to suit explicit pasture conditions and management objectives. Key grazing methods, for example, rest and recuperation, brush level management, water and shade arrangement, and legitimate fencing and foundation, ought to be executed to guarantee ideal pasture health, livestock execution, and environmental preservation. By taking on these practices, farmers can improve forage efficiency, increase conveying limits, and promote long-term sustainability in pasture-based livestock creation systems.

CHAPTER FIVE

PASTURE ESTABLISHMENT AND RENOVATION

Pasture establishment and renovation are significant cycles for keeping up with useful and healthy grazing lands for livestock. Whether you're starting another pasture or restoring a current one, appropriate preparation and execution are fundamental for progress.

Pasture Establishment

While laying out another pasture, a few key advances ought to be followed:

1. Soil Testing: Prior to establishing any grass or vegetable seeds, it's fundamental to conduct a dirt test. This will give you data on soil pH, nutrient

levels, and surface, assisting you with choosing the most appropriate plants for your pasture.

2. Site Readiness: Clear the site of any current vegetation, rocks, and flotsam and jetsam. Plowing or furrowing the dirt can assist with further developing seed-to-soil contact, promoting better seed germination.

3. Seed Determination: Pick grass and vegetable species that are appropriate to your climate, soil type, and expected use. Consider factors like dry season tolerance, grazing tolerance, and nutritional worth while choosing seeds.

4. Planting: Plant the seeds at the suggested profundity and rate,

guaranteeing even inclusion across the whole region. Roll or pack the dirt subsequent to planting to further develop seed-to-soil contact.

5. Initial Care: Water the recently established pasture routinely to energize germination and foundation. Screen for weeds and pests, and address any issues quickly to keep them from assuming control over the pasture.

Pasture Renovation

Renovating a current pasture includes reviving and working on its quality and efficiency.

1. Assessment: Assess the ongoing state of the pasture, searching for indications of soil compaction, weed

pervasion, and overgrazing. Distinguish the regions that need improvement.

2. Soil Testing: Like with pasture foundations, start with a dirt test to determine nutrient levels, pH, and other soil properties. This will direct your redesign endeavors.

3. Soil Air circulation: On the off chance that the dirt is compacted, circulate air through it to further develop water penetration, root development, and generally pasture health.

4. Weed Control: Execute a weed control program to wipe out obtrusive species that can outcompete helpful forage plants. This might include

mechanical expulsion, herbicide application, or a mix of strategies.

5. Overseeding: Broadcast new seeds over the current pasture to present superior grass and vegetable species. Make a point to utilize top-notch seeds that are appropriate to your particular requirements.

6. Fertilization: In light of soil test results, apply the fitting composts to address nutrient deficiencies and promote healthy plant development.

7. Grazing Management: Carry out a rotational grazing framework to forestall overgrazing and permit forage plants to recuperate and flourish. Change stocking rates and grazing periods in

light of pasture condition and development rates.

Maintenance

Normal maintenance is vital to keeping your pasture in ideal condition. This incorporates:

• Cutting to control weed development and support the new development of positive forage plants.

• Monitoring soil fertility and making fundamental changes through treatment.

• Rotational grazing to maintain harmony between pasture use and recuperation.

Note that laying out and redesigning pastures requires careful preparation, appropriate soil management, and continuous maintenance. By adhering to these rules and adjusting them to your particular circumstances and necessities, you can make and keep up with useful and reasonable grazing lands for your livestock.

WEED AND PEST CONTROL

Weed and pest control are essential parts of pasture management that assist with keeping up with the health, efficiency, and quality of the pasture. Compelling weed and pest control methodologies are fundamental for guaranteeing that the pasture can provide nutritious forage to livestock and supporting a different scope of plant species.

Weeds contend with helpful forage plants for nutrients, water, and daylight, which can diminish the general efficiency and quality of the pasture. A few normal weeds found in pastures incorporate thorns, dock, and ragwort. To control weeds, a few techniques can be utilized:

1. Mechanical Control: This includes truly eliminating weeds by hand pulling, cutting, or plowing. While this technique can be serious, it tends to be successful for controlling small weeds.

2. Chemical Control: Herbicides can be utilized to control weeds by applying

them directly to the plants or the dirt. It is fundamental to pick the right herbicide for the particular kinds of weeds present in the pasture and to adhere to the producer's directions carefully to guarantee a protected and successful application.

3. Cultural Control: Practices like legitimate grazing management, keeping up with ideal soil fertility, and promoting the development of advantageous forage species can assist with stifling weed development and lessening weed pervasions.

In addition to weeds, pests like bugs, rodents, and different animals can likewise present critical difficulties to pasture health and efficiency. Pests can harm forage plants, lessen forage quality, and send illnesses to livestock. To control pests in pastures, different methodologies can be utilized:

1. Biological Control: This includes utilizing natural enemies of pests, like ruthless bugs or parasitic nematodes, to control bothersome populations. Bringing gainful life forms into the pasture can assist with diminishing bug numbers without the utilization of chemical pesticides.

2. Chemical Control: Pesticides can be utilized to control pests by applying them directly to the plants or the dirt. It is vital to pick pesticides that are powerful against the objective pests and to adhere to the label guidelines

carefully to guarantee protected and mindful use.

3. Integrated Pest Management (IPM): IPM is an all-encompassing way to deal with bug control that involves numerous procedures, including biological, cultural, and chemical control techniques. By coordinating different control measures, IPM plans to lessen bug populations while minimizing the effect on the environment and non-target living beings.

Notwithstanding, powerful weed control and irritation control are fundamental parts of pasture management. By executing a blend of mechanical, chemical, cultural, biological, and coordinated pest management methodologies, landowners and pasture managers can keep up with healthy, useful, and different pastures that provide great forage to livestock. Legitimate weed and vermin control benefit the pasture as well as add to feasible agriculture practices and the general health and well-being of the livestock and the environment.

CHAPTER SIX
WATERING SYSTEMS

Watering systems play an essential role in pasture management. They guarantee that livestock receive spotless and adequate water, which is fundamental for their health and generally efficiency. Legitimate pasture management, combined with a successful watering system, can improve the quality of the pasture, upgrade animal execution, and enhance water use proficiency. We should go into the significance and types of watering systems utilized in pasture management.

Importance Of Watering Systems In Pasture Management

Water is a fundamental part of the well-being of livestock. Satisfactory water admission is important for digestion, nutrient retention, temperature guidelines, and generally the health of the animals. Insufficient water supply can prompt lack of hydration, diminished feed admission, unfortunate weight gain, and expanded vulnerability to infections.

Compelling pasture management expects to give ideal grazing conditions to livestock while keeping up with the health and sustainability of the pasture. A very well-planned watering framework guarantees that water is disseminated uniformly across the pasture, forestalling overgrazing and soil erosion. It additionally empowers even use of the pasture, taking into consideration better regrowth and delayed grazing periods.

Kinds Of Watering Systems

1. Ponds and Natural Water Sources:

A. Many pastures use lakes, streams, or natural springs as a water hotspot for livestock. While these sources can be cost-compelling, they might expect fencing to keep livestock from harming the water source and debasing the water.

2. Troughs and Tanks:

A. Portable or super-durable boxes and tanks can be utilized to give water to livestock in unambiguous grazing regions. These systems take into consideration controlled water circulation and can be moved to various areas depending on the situation to manage grazing and pasture usage.

3. Gravity-fed Systems:

A. Gravity-fed watering systems use the natural incline of the land to disperse water through pipelines to different areas inside the pasture. These systems are energy-efficient and require negligible maintenance, yet they may require a critical initial investment for establishment.

4. Pressurized Systems:

A. Pressurized watering systems use siphons to convey water to boxes or watering focuses all through the pasture. These systems provide predictable water pressure and can be modified to meet the particular necessities of the pasture and livestock.

Considerations For Picking A Watering System

While choosing a watering system for pasture management, a few elements ought to be thought of:

- Water Quality: Guarantee that the water source is perfect and liberated from impurities that could hurt the livestock or influence their health.

- Water Quantity: The system ought to be fit for giving adequate water to address the issues of the livestock, considering the quantity of animals, their size, and the climate conditions.

- System Productivity: Pick a system that limits water waste and promotes effective water use, for example, a trickle water system or low-stream watering gadgets.

- Cost and Maintenance: Assess the underlying cost of establishment, as well as progressing maintenance prerequisites, to decide the reasonableness and sustainability of the framework.

Be that as it may, watering systems are an essential part of pasture management. They play a crucial role in guaranteeing the health and prosperity of livestock, promoting ideal pasture use, and moderating water assets. By choosing and executing the right watering situation for your pasture, you can upgrade animal execution, further develop pasture quality, and accomplish maintainable and productive pasture management practices.

CHAPTER SEVEN

PASTURE MONITORING AND ASSESSMENT

Monitoring and assessing pasture is essential for keeping up with healthy grazing lands and streamlining livestock creation. Pastures give fundamental nutrients to animals and play a huge role in reasonable agriculture. Normal monitoring assists farmers with distinguishing issues early, executing opportune interventions, and settling on informed management choices. Here is a manual for pasture monitoring and assessment.

1. Visual Inspection

Start with a visual investigation of the pasture. Stroll through the area to check for indications of overgrazing, weed pervasion, exposed spots, or erosion. Search for changes in plant species creation and in general pasture health. Healthy pastures ought to have a different blend of grasses, legumes, and forbs with great ground cover.

2. Soil Testing

Soil testing is fundamental for understanding the nutrient levels and pH equilibrium of the dirt. Gather soil tests from various regions of the pasture and send them to a legitimate research facility for analysis. The outcomes will guide you in deciding the requirement for lime, manures, or other soil corrections to further develop pasture efficiency.

3. Forage Quality Appraisal

Assessing forage quality includes dissecting the nutritional content of the plants accessible for grazing. Utilize a pasture test or hand culling technique to gather forage. Present these examples to a lab for nutrient analysis, including protein, fiber, energy, and mineral content. Understanding forage quality aids in adjusting livestock diets and improving animal execution.

4. Grazing Management

Execute rotational grazing practices to forestall overgrazing and promote pasture recuperation. Share the pasture into smaller paddocks and turn livestock between them consistently. This strategy considers rest periods, permitting grass to regrow, and keeping up with pasture health.

5. Weed and Bug Control

Distinguish and control weeds and pests that can contend with positive forage species or damage livestock. Utilize coordinated pest management procedures, including mechanical expulsion, cultural practices, and specific herbicide application, to successfully manage weed and bug populations.

6. Water Source Assessment

Guarantee dependable access to perfect and satisfactory water hotspots for livestock. Survey the quality and quantity of existing water sources, like lakes, streams, or boxes, and make fundamental enhancements or establishments to address livestock issues.

7. Record Keeping

Keep up with point-by-point records of pasture conditions, management practices, and any noticed changes over the long run. Record the dates of soil tests, forage examinations, grazing rotations, and weed control activities. Keeping precise records assists in advancing, assessing management systems, and arranging future activities.

Note that pasture monitoring and evaluation are vital parts of maintainable pasture management. By directing normal visual examinations, soil testing, forage quality appraisals, and carrying out powerful grazing and weed control methodologies, farmers can keep up with healthy pastures, streamline livestock creation, and guarantee long-haul sustainability. Keep in mind that a very well-managed pasture benefits both the environment and the work of livestock producers.

THE END

www.ingramcontent.com/pod-product-compliance
Lightning Source LLC
Chambersburg PA
CBHW030050230526
45471CB00003B/1030